Forward for the Fenusian Experience

The reason for writing this forward are three fold: The First

is about proper lifestyle and diet. The body must have the

proper Nutrients, and other factors to have good health. If this

is not followed the result can be diasterious.Second, is how

two men who are successful in all phases of their life have a

void that needs to be filled. The final is an explanation of the

Title. The people of Fenusia call themselves Phenesians

Because the Group called the Phenesians originally colonized

Their Planet. The "experience" is the result of the people of

Earth and their relationship with the Fenusians.

Earl Hammers

San Jose, California 1-6-2015

The Fenusian Experience

A Change of Life Style

Chapter 1

It was a cold cloudy day. The winds were stirring the dust.
The trees and bushes quaking madly in the wind. The banners
On the buildings were blowing at 7+ flapping coefficient.
The Metropolis of Carolon was shrouded in the distance.

The buildings of the metropolis were hundreds of Meters
High. There were all kinds of Air Vehicles moving at various
Levels in and around the city. The buildings were various
Shapes including: Coils, Pyramids, Squares, Trapezoids, and
Circles in ascending levels. Above the clouds was the Cloud
Cities were the wealthier and more influential lived.

This was on the Planet of Fenusia; however the people
Referred to themselves as the Phenesians. This was due to the
fact that many millennia ago an advanced group from a
Galaxy far away, faced the destruction of their world through
A Supper Nova. A large group of lucky ones teleported to this
Planet of Fenusia, and their race survived. The planet of Fenusia
Orbited the far left star in Orien's belt. The labs were in a park

Like environment. The labs had living quarters were Terelis

Was rising.......

Terelis was a powerfully built man in his mid 50's. He was

A man of great knowledge and leadership quality's. He was in

charge of most of major projects at the labs, and he had armies

Of assistants. These projects had tremendous benefit for the

People of Fenusia. Some of these were: Time Travel, Every day

Life Improvement, and Ship Building and interplanetary

Commerce and travel. The most important however was his

greatest achievement; the development and discovery of

The Phenestrical Effect. This in brief was the following; It was a

combination of good life style, all things in moderation, and a

Chemical called Kurinite in the water and food to help the

people to live longer.

Terelis was wearing what could be described as a Red and

White Toga arrangement much like the Roman Senate. He had

A green belt around his middle heavy sandal like shoes. His wife

Valerias was coming in to bring him a cup of Java.

Valerias was a slim, quite attractive woman in her mid 40's.

She was wearing a long silk-like dress to her knees, a modest

Heel, nicely coffered hair in a light chestnut tied up in a bun. She had Pinkish Red Lipstick, and just a hint of Cologne. She said, "Good morning dear, did you sleep well?"

Terelis said, "How can I sleep? The Council refuses to budge And see reason?"

"Do they refuse to see your conclusion as to the cause of The plaque and the illness that has effected of our people" said, Valerias.

I do not want to return to the lifestyle of our ancestors. They were primitive and had limited technology; this was of A nuclear variety with endless wars and battles with silly names. We had mindless rabid Nationalism, which was exploited by A few; Also they would have us living in caves with limited Creature comforts, due to the displacement of battle. The only Good thing that we had was the Phenestrickle Effect, but it was Of little value because of the high death rate."

"I know that you have often spoken out against The Life Prolongation Program, which you have spoken out against My husband. You stated that due to your efforts; the people are Already living beyond 200 years, isn't that enough for one

People for goodness sake"said, Valerias.

I believe that Barnabus and Eleazar, and most of my staff
Are with me on this; however, the rest are adamantly opposed"
Said, Terelis.

Do you suspect a scandal or abuse of power in their refusal
To budge?" said, Valerias.

Yes I do, but I have no proof. But that is not the real reason
However" said, Terelis.

"What do you believe is the real reason for this attitude?"
Said, Valerias.

I believe that the real reason is they do not want to integrate
Into the Galaxy, and to share our knowledge with others. To
Have a cultural exchange and visitation of other people. They
Feel this would contaminate our world, absolutely ridiculous"
Said, Terelis.

Well dear, I think you should take a pause from your anguish
And have Breakfast. You always think better on a full stomach.
Look! Here come the children, let us not speak anymore of
This in front of them" said, Valerias.

"I do not think you need to concern your self with that,

I am sure they at least have a sense of what is going on. They
Have school friends, who have parents on the Council and the
Lab" said, Terelis.

The children came in, their names were Terelis Jr. and
Angelica. They were essentially in miniature the image
Of their parents.

"Good morning Mom and Dad we are ready for Breakfast,
And then on to school. Mom and Dad are we going to the
Mountains on our vacation?" said, Terelis Jr. and Angelica.

"Good morning children, we will discuss that later, let's
Have Breakfast" said, Terelis.

They sat down to Breakfast which was Bacon and Eggs,
With toast and Coffee. These were Yertal Eggs and Rondo
Beast. The Yertal was a large flying bird about the size of
An Eagle. The Rondo Beast was a pig-like creature raised for
Food but much larger.

The vacation was something the entire family had looked
Forward to. Terelis's busy schedule would mean the plans
Would have to put on hold. The climate on Fenusia was the
Following: It was very temperate. It was a nice blend between

Hot and cold. They were the third planet in their system, but a
Little further away from their sun, as compared to Earth. They
Had Oceans, seas, lakes, and Desert like areas. They had polar
Regions and Mountains as well. These extended to Thousands
Of meters in height. They were a vacation area were people
Would Hike, Camp and Fish. The mountains were where they
Wanted to go.

The home life was filled with every possible convenience
For the family. This was among other things an automated
Kitchen, and Bath, and Video Viewing System. They had
Robots to assist in all the activities of the household.
The family car consisted on an Air Vehicle, and a Pod System
For rapid transit for intra-city travel. They also had roads
Sprinkled about all over the planet, but were seldom used.

The Breakfast complete Terelis said,"I must make my
Departure. I am sorry children, but due to your Dad's busy
Schedule, we will have to put the vacation plans on hold for
Now. Do well in school children make your Daddy proud
Of you" said Terelis.

The children responded by saying "That's okay Dad we

Understand the situation better then you think. Don't work

To hard". They left.

Terelis said,"Good by dear" He gave her a long kiss on the

Lips and gave her a hug."I will see you tonight. Good luck with

Your meetings and speech" said Terelis.

Valerias said, "Thank you dear, good luck with your

meetings and other activities" said, Valerias.

Terelis got in the pod which would take him to the lab.

He knew it would be another day of wrangling, but he

Hoped it would not be so.

Chapter 2

Terelis arrived at the Council Champers and was

Immediately confronted by The Twelve, these were

The other members of the Council. Before this however

He was met by Baltisar. Baltisar was a man much like

Terelis. He was a man in his 50's. He like Terelis had a

Wife, her name was Miliseris. They lived in the Cloud City.

They had know each other for many years; since they were

Young men when they both served in the Military, and after

That had pursued scientific pursuits as well. They had for the

Last 50 or 60 years been engaged in the effort of improving

And informing the Phenesians on how to improve their lives.

They were also aware of the fact that not only were the people

Suffering from a plague, and general decay of health. They also

Discovered that many of the women were unable to have

Children, this had been going on for about the last 50 or 60

Years, and the population had plummeted.

Eleasar another friend of Terelis were friends of many years;

However in his case in was mostly through the Council. He lived

In the Cloud City as well. He also had a wife, her name was

Ruthias. He was responsible for most of the civic problems

Of Carolon and the Cloud City. Eleasar was more like a mentor;

As well as his Council duties he taught at the University on

Fenusia. That is were Terelis gained most of his Engineering

Knowledge. Terelis was his best student and was very proud

Of him; for his being in charge, and was the guiding star of the

Lab work and other projects. He bragged about this to his

Friends and the Council... He was getting too busy for the rigors

Of the work, but had in his younger years. The one unique thing

Was he had been married for over 100 years to Ruthias!

"Well Terelis are you ready to battle with those

immovable, intrangident and short sighted old dinosaurs"

Said, Baltisar.

"Well you know as well as I that they will not move even

If the roof fell on their heads. They cannot see the forest for

The trees" said Terelis.

Eleasar, Barnabus, and Baltisar said in unison, "You are our

last hope. You must convince them of a life-style change or

Our people will die out as a race."

They proceeded to the Council Chambers, where" the 12"

Were waiting. This is what the other members of the Council

Were referred to as. Terelis, Barnabus, Eleasar, and Baltisar

were Meet by Lorbal a dour faced sort, head of the Council. He

was Often referred to by Terelis as "Old Walrus Puss"

"Well Terelis are you sticking to your ridiculous idea, or are

you going to Listen to the voice of the people?" said Lorbal.

Terelis said,"I will continue as long as there is breath in this

Body of mine Lorbal. I believe as far as the voice of the people

Is concerned; they are not a short-sighted stubborn lot. I believe

They will support my position on the Phenestrickle Effect, as well As the intermix with the Galaxy", said Terelis.

Lorbal spoke, "We have our best Scientists and lay people Working on this problem. They feel as I it would contaminate Our people with a culture exchange".

Baltisar and Barnabus (pointing to the Council) "You do not Realize you have been polluting our world, and people. You do Not want to even see the good of ThePhenestrickle Effect or the Advantage of a culture exchange. You are destroying the health Of the people, and sterilizing our women with your Life -Prolongation project. A project we and Terelis oppose".

Terelis said, "I along with our team of Scientists and Technicians have achieved the best result of human Perfection by eating of the native foods and the water. We have achieved 500 years of progress in a tenth of that Of that time! We have achieved a 200+ year life span. Is That not enough for one people?"

Carmel Melonball spoke, "I for one, believe our people can Eat or drink anything they want. They can abuse their bodies For pleasure...and take a pill and solve the problem. I find it

Absolutely abhorrent to intermingle in the galaxy, and to be

Polluted by other cultures ideas. We do not need the

 Phenestrickle or you".

The other council members spoke in unison," We are a pure

Society and race, we will not be polluted and contaminated by

Outside interference."

Barnabus and Eleasar spoke, "We believe in our world too.

We do not believe you can obtain long life through artificial

Means and trash food and laziness. We do not believe in the

Scenario that you can reverse 40 or 50 years of bad living

By taking a pill. We also believe that we need to intermingle

In the galaxy; to not only get fresh ideas, but to possible

Help rejuvenate our women. There have been instances of

A culture exchange. The people who come very often make

The people happier and they enjoy the prospect of space

Travel. We can return the favor by helping them as well".

The other members of the Council spoke," You are

Dreamers, along with Baltisar and Terelis here... You

Believe the answer is out there. (pointing upward) "We

Do not believe we need outside help. We find the inter

Marring, absolutely abhorrent. We believe our people

Can partake of pleasure and not suffer from it. Our women

Are happy and have not complained about wanting different

men. The foods And water were never proved to be of

benefit".

Baltisar and Terelis spoke,"That is only because of our

Endless wars and deprivation. They had a primitive life-style

They were lucky to get a third of what we get".

"Enough! bickering is pointless gentlemen. Let's vote now.

All in favor of Terelis's plans say aih. (Four raised their hands)

All oppose (the rest of the Council raised their Hands).The

Decision is made….. the plan is not accepted", said Lorbal.

Baltisar, Eleasar, Barnabus, and Terelis spoke outside.

Baltisar and Eleasar said, "Well what will you do now

Terelis. The Council will not budge on either the culture

Exchange or the Phenestrickle effect"

Terelis said, "I have a plan. I have in my spare time built

A Time Ship. I will modify it and visit Earth of the Twentieth

Century, which is what they call our time, with a 50 year

Adjustment."

Barnabus spoke, "Why are you going to Earth? Are the reasons a fresh perspective on our problems. Do you believe The people will support   Extra-Terrestrials?"

Travelis said, "I have done considerable research on this With some of my assistants. I have found the people of Earth Are intrigued with space travel. They have had a short trip to Their moon. I believe for some, meeting Extra-Terrestrials An interesting experience".

Barnabus, Eleasar, and Baltisar said, "We wish you good luck To you on this endeavor. May the wind be at your back. Our Hearts go with you"

****

Terelis came home to his family. He gave his Wife a long kiss On the lips and gave her a hug. The Children came in and greeted their Father and said, "How was your day and the dealing with the council?"

Before he answered he brought the family into the dinning Room and said,"My dear family the Council refuses to adopt my Plan so I will modify my Time Ship and go to Earth".

The family which was Valerias, Terelis Jr. and Angelica said,

"We will miss you Father, come home soon, and may the wind

Be at your back".

With this...they had dinner; it was a very quit affair. After

Dinner he gave his children a peck on the cheek and they went

To bed. He gave his wife a long kiss on the lips and a hug. He

Then proceeded to the lab.

Terelis spent the next several weeks along with his assistants

Modifying the ship for high light speed. He spent a great deal of

time picking a crew for the trip. He also had Baltisar and

Barnabus. He picked Jerelis "The hottest pilot in four star

systems" He picked Miranda as ship's Engineer. He picked

Borealis as Cook. He picked Samelis as Navigator. He picked

The other Crew members for Maintenance and Security. The

Total including himself were twelve. There was no need for

more because much of the ship was automated. He had

Picked these people because he knew the quality of their work

And he had known them personally for many years. Terelis also

Picked Ravelis as his head of security; he could deal with any

Panic on the planet, stun the people and depart. He would be

Part of the landing party. Mavelis was a very studious fellow

He loved to study other cultures. He could help the group in

Their learning about the character of Earth People.

Chapter 3

After several weeks of extensive work the ship was ready.

Barnabus and Baltisar said,"We wish you good fortune in this

Endeavor and your voyage. We wish we could go with you

But we must watch for treachery and sabotage here on the

Planet".

The family came down to the launch area to see Terelis and

The departure of the ship. Valerias, Terelis Jr., and Angelica

Said, "Good luck Father may your trip bring a solution to

Both problems.

Terelis kissed his Wife good-by and said, "Good-by dear

I will be home as soon as our mission is complete. I love you

All very much" After giving his wife a long kiss and hug he

Went aboard the ship, and it left for Earth.

The planet of Fenusia orbited the far left star in Orien's

Belt. And due to the high light speed the trip only

Took a few months. When they arrived they began a

Functional orbit of said planet. They then proceeded to

Observe and record activities on Earth.

****

Paul Robert Jones was an Electronic Technician at Molecular

Video. They made among other things a DVD, CD, Radio combo

For large hotels and clubs. The customer in his room would

Simply go to the video screen and select from a menu what

Program or music or station he wanted to hear and it would

Be sent to his room. Paul's life had been a frustration, even

Though he had a good job and a nice place to live. His trouble

Was social, he was very shy. He was not bad looking, about

30, slim build, Dark Brown hair, and Blue eyes. He was very

Intelligent and read many books both Technical and Sci-Fi.

He also liked movies, dancing and a social drink, and smoke.

However, as a social butterfly he was not good, when in high

School, because he was shy.

After service in the Navy he went to college on the G.I.

Bill. He tried the bar scene in college, but found it boring

by the mindlessly boring music and the phonyness of

The women there. There only interest was fast cars and money.

Robert would get tongue tied too. He hard liked hard rock. The

Hard rock dens had only women who neither were nor really

interested In him either because of life-style. They did drugs

too, Robert Did not like that.

Although he believed in God he found the women somewhat

Unapproachable due to their desire for men who were mature

in The faith, whatever that meant. He found it difficult to talk to

them as well. He felt they had no right to make a judgment

Of a person. He has begun a good work in you will perfect it

In the day of Jesus!

He then joined a Square Dance Group. However, this

Proved Bad due to fact that he was not graceful. He could do

the Moves, but somewhat awkwardly. He was no Fred Astaire!

He tried conversation; although it was animated, and he lost

some of his shyness due to asking the girls, and talk etc.

He did not hit pay dirt. He could not do endless variations

Of the moves rounds and for other reasons he withdrew.

He was now about mid 40's in age after the dancing he

Became a huge Sci-Fi fan. He also became a Swing music

Fan, and collector of old radios and other collectibles. He was

Returning to his apartment when his friend Bob Smith showed

Up.

"So how's the rat race race"? said Bob

"Not so good and the rats are wining. I am happy with

The job it gives me a lot of experience in RF. Video, Audio,

Radio, And Mechanical; however my love life is not so hot. I

think it's Because I'm too shy and plane jane, no raze-ma-taze. I

have The what you see is what you get ideology".

Bob said,"Let's go the the Flea Market and search for

Treasures. That will be fun".

So they went because that was   the only pleasure they had.

At the beginning of the week Paul and Bob went back to

Work. They both worked in the Electronics field, which gave

them A sense of satisfaction. The supervisor Jimmy De-Castro

came to Paul and said, "I want to give you a commendation for

best test time on our products". He gathered the rest of the

test group together and said, "Gang, I would like you to know that one of your co-workers; Paul here, has the best test time

on our products, and due to this has helped the company a

great deal". There were congrats and applause from the others

of the Group. They had a pizza lunch and went back to work.

****

Bob was working at Bottom Mega-Systems and getting

Annoyed about the pinch-penny operation, and the general

Mess and untidiness of the place. He often said to Paul,

"I am the one that keeps that company going, and if it

Was not for me it would collapse. The owner is so disorganized.

He has me doing 10 things at once". However Bob was surprised

When one day the owner presented him with an award for a

Patent on improvement of the circuit. Bob loved R & D work,

And not so much the everyday trench and production work.

Paul was often annoyed about Bob's costant complaining

And suggested since he was not a nine-to-fiver, and he didn't

Particularly like the company that he get a different job. Bob

Would get equally annoyed at Paul so Paul said, "Very well

Let it be as it is!"

The following months and years went by and Paul and
Bob moved on with their lives. They continued in their
Jobs and going to a lot of Sci-Fi movies; they also continued
In Radio Repair and collecting old Radios and other treasures.
These included Catalogues, Post Cards, Newspapers and for
Bob all kinds of Visual and Audio communication and Recording
Devices of all types. This was all they had since they had no
Young women to share their life with.

Most of this happened before Terelis and the rest of the
Crew showed up. They had observed about a year of this
Activity by moving back in time and let the present time move
in A regular fashion. They did this for the whole planet; in that
Way they could observe the people for a long time, and
They could decide who best fit their needs. Paul and Bob
Were now in their fifties.

Terelis was up in the ship and observed these happenings,
He said to Miranda, "Well what do you think?
Do you like either one of them? Would you like to get to
Know either one of them?"

"They both are nice, and Bob in particular is very cute,

But seems somewhat withdrawn and anti-social. I would

Prefer Bob to Paul because he is more analytical and not

So spirititual. He see's things in Black and White, that's my

Slant", said Miranda.

Miranda, Jerelis, and Terelis came together for a meeting.

Terelis spoke, "I believe we can take these two Bob and Paul

Back to Fenusia with us. They both seem liberal enough for

This mission and they have no commitments or married."

Miranda said,"Yes, I believe they would be a good start.

I like both them; they are both painfully shy and never have

Had a girlfriend, I can help them with this. I have two distant

Cousins who I believe would be ideal for Paul; their names

Are Milinda and Valenzuela. Paul can decide, they both love

To Dance, lots of Make-Up and nice cloths. They both like to

Drink and Smoke occasionally. They are both very romantic

And I believe their personality's would fit Paul."

Jerelis said, "I like them too. Let's take them and maybe

A couple of dozen more, they can be specialists in various

Fields as well as lay people who want to travel. I will make

First contact with Bob and Paul."

Terelis said,"This should be a good start for our project.
Very well make first contact".

Chapter 4

Terelis remained on board, and Jerelis, Ravelis, and Mavelis
Made up the landing party. Ravelis was in charge of security,
And Mavelis was a specialist on Earth culture. He had spent the
Last 6 months studying all the idiosycraceys of Earth Life. They
Tele-ported down and Javelis made ready for the first-contact
With the unsuspecting Paul and Bob. The time of day was
evening, and Paul and Bob were on their way back to their
Car; after an evening at the movies.

A light appeared ahead as Paul and Bob approached the car.

"What is that light ahead? said Paul he was puzzled.

"Did you leave your car light on? I am always doing that",
Said Bob He was puzzled.

"I couldn't have. I have never done that", said Paul.

As they came closer they could see it was not the car.
They got in and proceeded home. It was at this time Ravelis
Made his presence known by saying, "I am from the planet

Of Fenusia, I will not harm you please come with me".

Paul was startled..."What the...What is this a robbery or

A prank?"

Bob was startled too he said,"We have just come from the

Movies, we have no money".

Ravelis said," I have not come for money please stop the car,

And I will explain."

"This is increadable",said Paul and Bob."Just like first contact,

In Star Trek a meeting of Extra-Terrestrials".

"What is this vehicle you are in?", said Ravelis He scanned it

And said,"hmm...primitive internal combustion engine burning

Partially consumed Hydra-Carbons. The fuel to produce this was

Called gasoline, and produced a by-product commonly refereed

To as smog. We had a similar vehicle on our planet but evolved

Above it".

Paul said, "What do you want of us? How is it you speak

Our language?"

"We have been orbiting your planet for quite some time

And have made a study of all your languages. The purpose of

our being here is for enter-racial marriage and to help

repopulate our world", said Ravelis.

"You mean to be used as breading stock, that's absolutely Abhorrent", said Paul and Bob.

"No there is more to it then that. You have in your makeup A possible solution to our problem. There is a second reason For this mission is a cultural exchange between our worlds. We can help you and you can help us", said Ravelis.

"I think it would be exciting to travel in space. This is even Better then the Space Program", said Paul and Bob."If we can Be of Help or service how can we refuse".

"I appreciate your cooperation; it is a matter of necessity" , said Ravelis. I will arrange for you and your vehicle to be tele-Ported up to our ship so prepare for departure", said Ravelis.

Paul and Bob were puzzled they asked, "How come if; as You say have been here several months, not been detected On our planet, we are not that primitive"?

Ravelis said,"We have a Cloaking Device which bends Light waves so we are rendered invisible. The power cost has been reduced so it is not a problem"

Paul and Bob said," What about our clothes and kitchen

Things, Furniture, and our Treasures and Toiletries, and
Our cars".

"Do not worry about that" , said Ravelis."We have made
Accommodations on our ship for your things. We will allow you
To make contact with your friends and relatives and supervisors
etc. We are not leaving immediately, we ask you not to reveal
The true nature for your departure. This may create panic or
any Other unseen event. We are doing this with all the people
we are taking. The people are all single, and have limited family
ties or other commitments that prevent them from leaving".

"Very well in that case we are ready to go", said Paul and
Bob.

After Ravelis had Paul and Bob beamed up to the ship, he
Along with Javelis and Mavelis had similar encounters with the
Other people they met. In fact with the aspect of teleporting
From place to place, and knowing exactly were the people were
Were; it only took a couple of hours to get all the people they
Needed. They took the two dozen that were needed and they
Were all teleported up to the ship.

When all the people were aboard Terelis had them assemble

In a gymnasium like area. Terelis addressed the group"Ladies

And Gentlemen you have been brought here to help our planet,

To perhaps help yourselves as well. We know that you are right

on the verge of the beginning of a large amount of research,

However you lack money. We do not have that problem. In fact,

We can provide unlimited resources in that. You can make

Tremendous progress in your chosen fields here on Earth. We

Have a problem because of an abuse of progress. We have

Created a plague on our planet. The women have become sterol

As well. You may have the critical element we need. What is

more important is the need of fresh blood and ideas for

Our world. We can share our knowledge with you, so you

Do not make the same mistakes we did. We need a real

Culture exchange to help our world integrate into the

Galaxy and help our world of Fenusia. For the record

That is where we are going, to a planet circling the third left

Star in Orien's belt. We may not need you just for breeding

Stock as many fear. Will you help us?"

The group responded in unison,"We are somewhat

Primitive In our technology; however, we are not advanced as much as You but we are eager and anxious to help You!"

Terelis said,"We are indeed grateful for your help, and we Can make ready for departure". This having been said the left Orbit and headed to Fenusia.

Chapter 5

There arrival at Fenusia was a mixture of good and bad. Valerias was delighted, and she hugged her husband. She was Wearing a soft silky top. She had a scarf around her neck with a Pendent and the two pieces were fastened in the center. She Had Pinkish-Red Lipstick, nicely coiffured hair, and White Skirt And Heel. She gave him a caress and long kiss full on the lips. She said, "I have labored hard for you while you were gone. I have spoken at many meetings in the park, culture settings, And others are telling the people about your plan. The plan should At least be tried. The people of Carolon and the Cloud City At massive rallies have cheered, and think the plan should Be tried. The children, bless them have also spoke at their School with their friends, and their parents and most agree. I am anxious to meet the people you brought back", said

Valerias.

Terelis said,"Thank you my good and faithful wife. I could

Not be happier at this news. All of the people we brought

Back arte anxious to help us. They are somewhat over whelmed

And happy to be on another planet and to travel in space. It is a

Big deal for them. I think you will find they are very easy to

know, and to work with".

Terelis was somewhat sad and pensive now....

"I suppose the 12 and others have been equally busy

Trying to defeat my plan?"

Valerias said, yes, particulary Lorbal. He has as many people

On his side, that they say that they are the voice of the people

And the planet."

"Enough!" said Terelis I want you to meet the people

We brought from Earth".

The people of Earth were deeply moved by the city, and the

sediment shown them. They were meet by the opposite sex

Research people who said," We are glad you are here. We hope

Together we can solve the crisis at hand".

The people of Earth responded thus, "We the people of

Earth we are grateful for this opportunity to help you in any way

We can".

The people of Fenusia were humanoid in appearance, but

had some interesting differences from Earth they were

the following: The Men and Women both had large full lips.

Both the Women and the Men had large eyes, and were usually

of different colors. The hair ranged from extremely short and

Fuzzy for the men and long and curly for the women in various

lengths. The color was from Blonde to Black to Chesnutt color.

Their build was usually very muscular and thin for the men and

the same for the women but not muscular. They had for several

Millennia been very vital physically, but in the last 50 or 60

years had declined and the women had become sterile. The

population plummeted. They had their own special

Electronically produced music more like a combo of Synthesizer

and Computer. They loved the arts, and were extremely good

at adapting language and incorporating very quickly any

contacts they made. This is why they learned about Earth's

culture so quickly and adapted It, this applied to the more

liberal of the Planet. People like Lorbal and the council felt it contaminated their world to have others there. The Men were immature and arrogant towards the Women, and the Women loved romance and the arts. This is why Milinda, Miranda, and Valenzuela liked Paul and Bob so much and the other women of Fenusia did as well. When they met the other Men Scientists and specialized Technicians. They also had a curve upward on the top of the ear. They also had large lobes below the main Ear, ideal for piercing and earrings which almost all of the Women of Fenusia had.

****

Terelis said, "Before you meet the other people, there are Two people I particularly want you to meet. They are Paul and Bob and I want to help them; You remember those two cousins Of ours who seem unhappy with the men we introduce them To, they might like them."

Paul and Bob came forward and they happened to be wearing the outfits they had on when they were teleported aboard the ship along with their treasures.

This was; Blue Levis, and cotton Diamond patterned Shirts,

White Socks and Brown Walking Shoes. Bob had on a similar

Outfit.Paul's hair was nicely combed, and Bob's was asque and

some what flat.

Paul extended his hand and said," What a thrill it was to

travel in space, and meeting Extra-Terrestrials just like in

The movies. Your city is beautiful and I also saw your anti-

Gravity city are very unique from anything I have ever seen.

If you don't mind my saying so, I think you are very attractive

In dress and appearance, it is a pleasure to meet you."

Terelis said,"My dear I think you have made a conquest".

Valerias said,"Thank you that is the nicest compliment

I have had. I am surprised some girl did not grab you off

Years ago. Our two cousins will be delighted to meet you.

You are quiet and unassuming they both like that". She laughed

Before she said this."From what my husband tells me you are

Very intelligent, somewhat shy, you should fit in on our planet.

Bob came forward and said,"I enjoyed the trip here, but

Didn't see much. It's not like in the movies. I have a more

Realalistic view of space travel".

"My husband mentioned that you were the more serious

One. You are not as romantic about space travel as in Sci-Fi

Movies, but more analytical about life and other things

Every thing needs to have prove .Just wait till Miranda meets

You. You would be like two peas in a pod. She is my husbands

Chief Engineer", said Valerias."All things need to be proved.

Every thing in Black and White that's our Miranda".

"I have not had the pleasure ", said Bob.

Paul said I love a Dance, nice Clothes. I like women who

Have Makeup nicely coiffured hair, cologne, they are good

Conversationalists are fun to be with. They like all kinds of the

performing arts, and romance and candlelight dinners

And other things as well. Do you have anyone like that?"

"Yes", said Valerias"My husband and I have two cousins

I'm sure you would like, and I'm sure they would like you.

You're so Wonderfully shy and unassuming. They are Milinda

And Valenzuela, But, before that happens my husband wants to

Talk to the whole group.

Terelis said, "Welcome to Fenusia! The purpose of our

Mission here is the following: First to blend into our culture as

Much as possible. You may find after a time you wish to remain.

Second to have our Scientists examine you to determine a

Solution to our problems here. Third to incorporate this into

A solution, that is palatable to both. Fourth if this works, we

Bring back more of your people to either intermarry or to

Provide the critical ingredients we need. Fifth as an added

bonus a culture exchange between our worlds. We know that

Your planet is in a very troubling aspect now; we can help you

With this, we can keep you from making the mistakes we made.

Since you have had only relatively primitive space flight

Program we can help you with this too. The benefit we get out

Of this is that we can learn about other cultures; we need to

Integrate into the Galaxy. You would help us save our culture

Which we would be immensely grateful"

Paul and Bob said, "We would like to see the sights of

Carolon. The first things Earth people do are sight seekers".

Terelis said, "That can be arranged, in fact Milinda and

Valenzuela will be your personal guides. In fact arrangements

have been made for all the Earth people to have guides around

Our capital of Carolon. The guides will be the opposite sex of the

People being guided; so as much romance and friendship can

Be developed. This was one of the reasons we took all single

People. The disruption of family life would be at a minimum.

The family life aspect is important to we Phenesians as well"

Valerias and Terelis took Paul and Bob to their living

Quarters which was a short distance from the plaza, when they

Had arrived..

Valerias said, "I would like you to meet our children they

Are Terelis Jr. and Angelica. Children come in a moment and

Meet our guests"

The children came in and bowed slightly and said,"We

Are pleased to meet you. We hope you can help our Dad

And our world. The other children wish this as well".

Paul and Bob said," We are glad to meet you as well. You

Seem remarkably poised and mature for your age. We are

Impressed and pleased by this".

Terelis Jr. and Angelica said,"Thank you; the reason for this

Is from our Mom and Dad, they have given us the spirit that you

Speak of." When the children meet the other people of Earth

They were equally impressed.

Valerias then said, "I am amazed by this vehicle that you

brought back. My husband tells me that it burns partially

consumed Hydro-Carbons. It was propelled by a fuel called

Gasoline, and produced a by product called smog".

Terelis said,"ahh yes I can remember as a small lad, my

Father speaking of such a vehicle. We have not used them for

almost a hundred years. We have air vehicles which operate by

a Solar and Electric combo. They fly at various levels and on

roads, but we seldom use them. We have an extensive pod

System for intra and inter city travel".

Bob said, "They are called cars or automobiles. We attach

much of our culture and mobility to them. For nostalgia and

Pride we call the older ones"Detroit Iron" named after one of

the cities that made them".

Valerias said, "I find it interesting about your planet, that you

place such a mystique about them, for a device of necessity.

We have a surprise for you. We wish you to meet your guides

To our city and planet; Girls come in and meet our guests."

Miranda, Valenzuela and Miranda entered and extended

their hands to Bob and Paul and said," We are glad you are

Hereto not only help but to give you a tour of our world". They gave Them a hug after saying this.

Miranda walked over to Paul's display vehicle and said," Your Culture places a great deal of necessity, status, and a certain amount of romance to these vehicles"

"Particularly on a warm summer evening!!" said Milinda And Valenzuela. They grabbed their middles and shock Themselves as they said this. "What a romantic interlude".

Miranda walked over to Bob's display of Radios and said, "These are interesting too; what did you say they were? Radios I believe you said? The construction is all wood in some cases and others heavy plastic and well made.

Paul and Bob said, "We like to collect and repair them".

"Hmmm... Amplitude and Frequency Modulation on carrier Wave signal. I can remember in one of my communication Classes our instructor talked about them. They were extremely popular with the people of our planet about 100 years ago."

"Oh for peat sake Miranda! We just love the music that came from them. This was ideal for Romance and Dancing, Dancing, Dancing". , said Milinda and Valenzuela. They griped their middle and shock themselves as they said this.

"We will see you in the morning" said, Miranda, Milinda,
And Valenzuela, and they left.

"We will see you in the morning as well" said, Valerias and
Terelis "We will have our robots guide you to your quarters
where all your treasures are stashed"

Paul and Bob retired to their rooms, to see what the next
day would bring to them. They were impressed and somewhat
taken aback by the whole layout of the planet. They
vowed that they would adapt.

Chapter 6

The first few days were more or less for orientation. The
Phenesians and the Earthmen had many representatives for all
kinds of Scientific disciplines. The two parties were sharing
information with their colleagues, and found the Earthmen
remarkably helpful. They did full physiological and physical
Aspects of both. The Earthmen and the Phenesians were using
this As A starting point. One day Terelis came to Bob and Paul

and said, "There is someone I like you to meet formally since

the last time was somewhat brief .My wife and I know them,

they are cousins of ours, and the third is my ship's Engineer on

Our trip to Earth. They are Milinda, Miranda, and Valenzuela."

The three women were as follows: Miranda was the ships

Engineer. She had a trim figure with long Blond hair. She had

one Blue eye and one Green eye. She had full soft lips and

Smooth skin. She had modest eyelashes and no makeup.

When she was working she would usually ware coveralls;

But when not working she did like very stylish Pantsuits,

And sometimes Blouse and long skirts. She usually wore a low

Heal with these. She liked to wear with her skirts knee high

Boots. She had the Veronica Lake swoop in front on her hair.

Melinda on the other hand loved soft frilly things. She liked

Lots of lace, she like pearl necklaces, but no other Jewelry

except Maybe an occasional ring. She also liked long silky

Evening gowns, With Black High heel shoes and Pinkish-Red or

Red Lipstick. If she really wanted to kick up her heels, she

She also liked a

hint of Cologne. She enjoyed dancing and romancing, and

Candle light dinners. She would also go to plays and concerts

on Fenusia. She sometimes went out with men but usually

Friends of hers From the office where she worked as an

Executive Secretary For a large manufacturing company. She

was well read and very Intelligent. She had a very friendly

outgoing Personality, but Some what subdued. She had long

Curly Brown Shag hair, like Meryl Streep, or Glenn Close.

Valenzuela was between them both. She had long curly Black

Shoulder length hair. She had long eyelashes and large Brown

Eyes. She had large full lips which usually had Purplish- Red or

Pinkish- Red lipstick. She had extremely soft smooth skin, and

an Attractive face. She was pleasingly plump unlike the other

Two. She loved dancing the most, and lived for this. She had a

Serious Side too, she was very spiritual and often went to

Church. She liked the Arts too but not as much. She had a

Secret crush on Paul, but shortly there after fell madly in love

with Mike Boyce who was the pastor on the trip out from

Earth. She like Miranda and Milinda liked very stylish clothes

and wore Spike Heels and a touch of cologne. She like Miranda,

and Milinda had nicely Coiffured hair. She was a Teacher of

Language and Cultural doings. She had a very bubbly

Personality, and very outgoing socially. She was truly a free

Spirit socially .She   had a Hedi Lamar Hair style.

Paul came forward and said, "I am pleased to meet you and

From what Terelis has told me, you are women after my own

Heart".

Valenzuela and Melinda said, "We understand you love to

dance, and are romantic, and somewhat old-fashioned in your

Own way. We hear you love old movies, both Dramas and

Sci-Fi. We like old movies too but not so much Sci-Fi with

Space ships and such; but more for romance and love."

Paul then said," You are both very attractive and I would

love to have a social encounter with you; but we must put that

on hold till after the crisis."

Valenzula then said, "Where's your friend Bob? I am sure

That Miranda would love to meet him?"

Melinda then said, "If you look over yonder (finger pointing)

I think she has found Daniel Boone"

They were deeply engaged in conversation, and lost to the

World.

Miranda had walked over to Bob who had been looking at His Radios and checking his Car and she began talking to him it Wasn't long before she said, "Bob I just love your analytical Mind. You are not particularly spiritual and I like that too. I am so tired of the phony and plastic men here. I mean they Are all right in their own fashion; however, they have no real discipline in their life. They just want to party all the time and have a good time. I like your hair in a somewhat askew style, Unlike your friend Paul. They are also very arrogant, conceited, and self-righteous, and contemptuous of women's feelings. That is a real turnoff for me".

Bob then said," Thank you Miranda for that. I am the serious One, all my patterns of thought are Black and White. Every thing needs proof for me. Paul and I have been friends for many Years, and many of the same interests. You are what we call a kindred spirit and I'm grateful and glad that we meet and can be friends".

Miranda reached up and caressed Bob's face and said," I believe in time we can be more then just friends. I really think that I love you Bob, I have never had this feeling about anyone before. With me it has always been work'"

Bob then said,"I believe for the moment we will have to put Any kind of social encounter on hold till after the crisis, like Terelis said.

Miranda shock her head and said,"Your right I forgot the logical part of this meeting, you are right I forgot myself in the emotion of the moment."

Miranda then said, "We can do something this evening (With a twinkle in her eye)"I think you would get a kick out of. This is in honor of your coming I have built from scratch a complete box T.V. with A.M and F.M. radio with Cobra From vintage parts from scratch! "

"Bob then said,"I would love to see that and watch some Vintage shows as well. I would like Paul to see your tube as well".

Miranda said, "Let's leave Paul out of this for the moment". Miranda took Bob's hand and led him to were the tube was. She caressed his face, and put her arms around him and gave him a long kiss on the lips. She was delighted to show off her Engineering skills, as well as affection. This was particularly true in this instant, because she had never had any real feelings or interest in the men she had encountered. They were in her mind "a real turnoff".

Melinda and Valenzuela said, to Paul "We have studied your Culture, after Terelis told us about you. This had happened After he had studied your planet for many months while orbiting your planet. We have a film preservation society For our movies and so we decided to do yours as well. We focused particularly on what you call the golden Age of Hollywood from the 1930's to the 1950's but have Representatives of all eras. We have records of the movie Palaces and have created them here on Fenusia. This is so You can feel at home on our world. The people here on Fenusia are coming in small groups, but now that you are here This will increase. We have a collection of vinyl as well, everything from Rock to Bach! You as a technical guy may be interested in the way we incorporate the data. We use Nemesis- Blue and Ultra-Dark memories. In addition there is 100 percent Visual and audio quality. No compression, movie quality video. We have reproduced the Movie Palaces and Big

-Band DanceHalls, but we have made the same for Rock and Classical."

After this Paul talked with Milinda and Valenzuela about All kinds of subjects. They particularly talked about Dancing and Romance. They were both thoroughly delighted with Paul.

Paul said, "I am happy and it pleases me that you both have have made such an effort in my behalf I….."

Melinda and Valenzuela then said, "we are delighted that you are here; you are like a breath of fresh air. You are delightfully Unassuming and shy we like that"

At the end of the evening they said to Terelis and Valerias, "We are glad that you brought us these two men and look forward to going out with them particularly Paul".

Paul and Bob had already left and Terelis said," I am afraid your plans will have to be put on hold for the moment" The Three women Melinda, Miranda, and Valenzuela seemed disappointed but took it in stride. Miranda then left.

Valerias then said,"which one do you think likes the other, And do either of you have feelings for either?" (With a twinkle In her eye)

Valenzuela said," I really like Paul, Milinda I think he likes you

more then me, I am a little disappointed but there is always

hope. He sure is cute though".

Melinda said, "I wouldn't underestimate your charms, let's

wait till after the crisis ,and he can date both of us and make a

decision." I agree as well." Said Valenzuela.

Milinda and Valenzuela then said," I think Miranda is really

Smitten with Bob, I haven't seen them for a couple of hours.

She's conquered and he's Sherman marching through". They

both said goodnight and left.

Valerias and Terelis were alone now and she said," Well my

Husband I think we are well on the way to playing matchmaker.

Terelis said," You are right my dear, I'm so glad that we can

help these two lonely men. They have been an immense help

To me and my staff with there help".

They both went to bed at this time, both very happy at the

turn of events. From this point on Paul and Bob referred to

Milinda, Miranda, and Valenzula as "the big 3".

****

The Research Group from Earth after several months had completed their initial research. They met in a conference room Given them by Terelis for their meetings to be held. Dave Jones and Dick Thomas spoke first, "After extensive studies and Investigations we have found that the main difference between Us and the Phenesians is a missing critical gene for reproduction And also for physical growth and vitality".

Cindy Smith and Edith Evens said,"We have a larger lung capacity then theirs, there's seems to have shrunk. They have no Appendix. There Liver, Pancreas, and Prostate and uterus are larger then ours".

Carol Turney and John Sherman said, "They have an extra Organ in their chest, which seems to assimilate and dispense Energy and can promote growth".

Jim Evens and Mark James said, "They have a circulatory System Heart, lungs, and thyroid much stronger then ours."

Vallary Sherman and Dick Tomas said, "It seems that they are victims of their own technology, and abuse of some

Artificial ingredient that has affected their entire physical

Makeup and energy. It appears they are healthy but seem to

lack something".

The other Scientists and Technicians who had not spoken

Said in unison,"Because of the combo of all these factors, and

perhaps more they have been victims of their own success.

They are too close to the problem and can't see the forest for

The trees. We must inform our colleagues of our findings".

Upon meeting with the Phenesians together they said,"

We are truly amazed in your findings and are grateful".

Thomeris and Markeris said,"We are grateful for this

Conclusion, and it seems logical"

Jameseris and Vickyeris said," We had come to the same

Conclusion as well; we could scarcely believe it, and we did a

Double check".

Susanis and Robertois said, "In addition to what you have

stated, we have found that the people have not been eating

The native foods and water for the last 60 or 70 years here on

Phenesis."

Bettyis and Davidis said," The deteriation in the body has

Been caused by a lack of Kurinite in the blood stream. Our People discovered this centuries ago and it helped with our overall Health and vitality. Terelis who is the man we all work for had tried for the last 50 or 60 years to convince the people and The Council of the effects of being lazy and not eating the Native Foods and water. The public wanted convenience and felt they could reverse 50years of excessive drinking, smoking, and eating and endless self indulgence. They felt they could reverse this with the pill. Terelis our leader and colleague were deeply opposed to the Life Prolongation Project and had fought a losing Battle. We feel that there is a scandal involved but we can't prove it."

They all said in unison, "The people are destroying themselves with this pill. The plague is the prove."

Rebecaeris and Josephus said, "We must get this to the Council and particularly Terelis immediately. We hill hold on the Suspicion for the moment".

Upon meeting with Terelis he said,"I am truly grateful for Your findings, and that an independent group have verified My findings of years ago. I can have some hope for our world.

I believe that Lorbal and the Council and some very powerful People are behind this". The group stated that they had the Same suspicion but could not prove it. Terelis said, "I will go the Council tomorrow and because of the urgency of the matter I will request that the City of Carolon and the Cloud City as well as the outlying area representatives are there".

After they left Terelis had a deep sense of satisfaction That a outside condition, and totally neutral group had Confirmed what he had been saying. After this he went home. Valerias was there and gave him a big hug and kiss. She put her Arms around him and caressed his face after which she said, "My husband I am so glad of the news you have told me. I am happy for our people, but particularly for our children".

The children came in and saw their parents embarrassing And said,"We Terelis Jr. and Angelica are so proud of you Mom And Dad. Our friends are happy as well. We are glad for the Efforts of all the people who have helped on this project".

Valerias said,"How are you going to convince the Council? They would never approve of matting with the Earthers. I have Seen and I have heard a great deal of affection, particularly the

Women, for the Men. The Council would never approve of the Plan, they don't even like the idea of them being here."

"Yes I know, said Terelis."We must try however, they must Be made to realize what is at stake for the future of our world. I believe if we don't adapt this plan and get back to basics, we Could die out as a race in less then 50 years".

Valerias and the children were horrified at this news and said," you must prevail Father".

"I am not afraid of the Council now and the 12.I believe with The other representatives of our world plus Carolon and the Cloud City that will be there they will provide additional force against the Council, but it will be a hard sell".

Terelis was pensive for a moment then said," As you have Stated I have noticed a great deal of love and affection that the Women in particular have shown to the Earthmen. I believe if we are successful with this project, that we can have a real Culture exchange between Earth and Fenusia."

Valerias said,"Upon the morrow you shall go to the Council With your findings and proposal for our planet. I wish you good Fortune in this endeavor". They embraced and gave each other

A big hug and kiss.

Chapter 7

The meeting of the Council produced the results expected.
Terelis, Barnabus, Baltisar, and Eleasar were there to represent
A push for the adoption of Terelis's plan, and the integration of
the planet with the rest of the Galaxy which was Earth for
the moment.

Lorbal was furious with rage and said, "I have never heard
such a preposterous proposal. I for one will never submit to this
humiliation" He said this in front of the Council in the chamber.

Barnabus and Eleasar said, "We are a small population on
this planet, and getting smaller. If this problem is not addressed
It will be the end of our race. Our experimenting with The Life
Prolongation Program, Which was disapproved by Terelis,
Baltisar, Eleasar and myself as we predicted as had disasterious
result. Our improper diet, excessive drinking and smoking and
endless self indulgence have weakened our race and sterilized
Our women. The only way to save our world is a combo of inter-
Marriage and more important, change of life style. There is

Record of many worlds were this has helped".

Morlock and Carmel Melonball spoke, "We are livid with rage. We do not want to contaminate our world by inter-marriage with outsiders. This is absolutely abhorrent to me."

Baltisar and Terelis spoke next, "It wouldn't be just one way. We could help them to avoid the mistakes we made. We could, With the proper blend of technology give them a life of comfort And pleasure. They could give us a breath of fresh air with their Unassuming ways and new culture. We could help improve their Space flight capability so they could get out in the galaxy. I have Found they are immensely curious about the universe".

The other members of the Council were so intimated by Lorbal, Carmel Melonball, and Bayner Mc-Connell that they Very rarely spoke their mind except to say, "We will tell the people of your plan and they will reject it."

"Enough! said Eleasar, Barnabus, and Baltisar,"We will do the same. We will campagne planet-wide for the proposal, and we Will see who wins".

Terelis said, "I believe you are wrong about the people. We Collectively believe the people are at least liberal enough to at

Least an exchange with the Earthmen."

Morlock and Lorbal had always been a thorn in the side of Terelis. Morlock had been Terelis's assistant but rebelled about The Life Prolongation Project. Morlock could see the prospect for tremendous profit from the project; by selling the public a "Bill of goods" about the pills, and they would have absolutely little or no benefit. Terelis was furious about this, and had him Expelled from the lab. He had tremendous technical ability And was very jealous of Terelis's success.

Carmel Mellon Ball and Lorbal were for the most part, played the role of the stubborn beaucrate. They had no real Technical knowledge, but knew power and how to control it. they could employ knowledgeable people to swing their way with the Proper amount of favors and bribes. They were always at swords point with Terelis about civic matters on Fenusia, Carolon and the Cloud City. The other members of the Council did not like Terelis's proposal either, but they more or less followed Lorbal's lead Most of the members maybe in their Minds had toyed with the idea of Terelis.

****

So the battle began in earnest...... Terelis, Barnabus, Eleasar,

and Baltisar; also the wives Ruthias and Miliseris, along with all

of their friends began to speak at massive stadiums and parks

On Fenusia.

Carmel Melonball, Morlock, and particularly Lorbal spoke

at Huge rallies as well to convince the people not to accept the

proposal of Terelis and his group. After many speeches both for

And against the proposal an election was held to decide the

matter. The results were that the people, of which the majority

Decided to try the plan for 6 months. Terelis was delighted with

This news. He told his wife about it sand she was delighted as

Well. When the children heard about it they said to Terelis

And Valerias." We are thoroughly pleased with the result, and

this means that we along with our friends and the people of

our world have a hopeful future. We are proud of you Dad for

being Successful in this matter".

Valerias said, "I am happy for our people for two reasons.

One a hopeful future for our women and children. The second

it means the women particularly can marry the Earthmen. I have seen a great deal of affectations there"

Terelis said, "It is a great victory but we must be prudent and cautious. There is no telling what Lorbal and the council will do".

Meanwhile on an another part of Fenusia.....

Morlock, Lorbal, and Carmel Melonball came together for a meeting. They were with the other members of the Council, plus many of their minions. Carmel spoke, "I am furious with the result of the election, this really spoils our plans".

Carmel spoke, "We had plans for other Star systems, in addition To our own to conquer them and to poison their people with our pills, just like here. This really puts a monkey wrench in our plans".

Cavelis and Mavelis spoke, "We will continue here on our own planet for the moment. The people are totally unaware of What we are doing."

Morlock spoke, "I for one will not give up our huge profits, and the cozy relationship with the regulators and such, through bribes and sexual favors and goods. We have been selling our pills and the people are totally unaware they are useless". He had a tremendous burst of laughter after this.

"Enough! spoke Lorbal "We know the people are gullible and buy and believe almost anything. We have support of many people, who wish power and profit. We have support of many Scientists who have been convinced to support us, This includes In this group lay people as well. We will become very rich for many years to come with our plan.

So for several months the plan of Morlock's worked due to delays in incorporating Terelis's plan. The lobby was so strong that despite the election the propaganda was so intense that the people were afraid to break away. Even the people who had voted against the Council were helpless against them. Terelis and his group tried to break this grip on power, but failed until as luck would have it a break....

Some of the members of the Council and also very influential Lay people unbeknown by the 12; that despite the fact they were not in favor of Terelis's plan, they felt the deliberate Poisoning of the people for profit was abhorrent to them. They informed Terelis of same, and he was livid with rage.

Terelis said, "I cannot believe that some of our own people would deliberately poison and mislead our people for profit. They must be stopped". He pounded his fist on the table as he said this.

Valerias said, "What would you do to solve the problem my Husband?"

Terelis said, "I will publically denounce the Council of the 12 and the others involved with this evil plan, and have them banned from Fenusia. I have never liked Morlock, Carmel Melonball or Lorbal. They have been a constant thorn in my side: ever since he was disgraced and had to be dismissed from the council. The three of them have tried to stop every thing I have tried to do. I believe Barnabus, Eleasar, and Baltisar will agree with me. I say good riddance!"

Valerias was somewhat skeptical and said," Do you think Dear that they the people will believe you. You have had trouble with them in the past. People just will not listen to you".

I believe when the people realize what fools they have been They will believe me. I have Barnabus, Eleasar, and Baltisar on my side as well".

So he did this; Morlock, Carmel Melonball, Lorbal and all the other minions of Lorbus were banned. Terelis was delighted when the people had proof of the scheme they rallied to Terelis. They were angry they had been deceived.

The men were banned to a moon called Fusiliers which was a Penal Moon of Fenusia.

****

With the trouble makers out of the way the plan was implemented. Paul and Melinda had several dates and were married on Fenusia. He was promoted to special assistant to Terelis on scientific projects. Paul went to school on Fenusia to Get the necessary degree, it took four years. He continued to work on his radios and he took Melinda for rides in his car. They were hardly any vehicles so they drew huge crowds Melinda loved this. They did a lot of dancing and dinners and The Arts on Fenusia.

Bob took Miranda out several times. These were very

Intense lectures and scientific in nature. They were Space Flight Time Travel along with other scientific discussions. They fell more deeply in love and truly enjoyed each others company. Bob was promoted to assistant Engineer on Miranda's ship. After 4 years of college Bob was an able assistant. They were married by the skipper of the ship and spent their honeymoon on a Geological and Archeology Expedition. They had been married on the ship because it seemed like a second home to them. They took a trip to Earth to study Earths physical makeup and all of the scientific problems including population and tried to resolve them. This was much like the original trip. They studied Geology, Weather, The Growing Cycle, Archeology, Environmental, Food Growing and the entire field o Scientific Endeavor. Bob found it interesting to return to Earth and tell people about the opportunities for Research and knowledge on Fenusia.

Valenzuela continued her dancing; after a short and Unsuccessful attempt for Paul she met Mike Boyce. He had come from Earth to share his beliefs with the people of Fenusia. He found that they were remarkably eager to embrace the faith. Mike had a very bubbly and vibrant

Personality. Valenzuela found him delightfully unassuming and

a pleasure to be with. Valenzuela fell madly in love with Mike

and they were married in a big church wedding which was his!

He was not the kind of Man who wore his religion on his sleeve.

He enjoyed dancing and going to the arts on Fenusia. They

became very close friends of Paul and Milinda.

After the plan was enacted several more trips were made

To Earth. Travelis was in charge of this project but did not go

himself. The people of Fenusia found the people delightfully

Unassuming and a breath of fresh air. The women many of

who found the men to be arrogant and uncomplimentary

Returned to Earth. This unfortunate situation changed when the

Women were going with the Earthmen. The men decided to

Straighten up and fly right; so after a short time the women

returned to Fenusia. The culture exchange was a tremendous

Success and Terelis was justly proud.

The ones who refused to participate or change were

Winnowed out. Terelis discontinued the Life Prolongation

Project and its facilities were used for other projects.

One of Terelis's jobs was to inform the people about the Natural sources of food and water on the planet. Barnabus And Eleasar continued to help Terelis in all of the projects. Valerias and Miliseris were in an active life of lecturing and raising Their children. She was also active in support of her Husband in speeches. She also continued to teach at the University. She supported her husband with accepting the life span of the people a little over 200 years for goodness sake!

One of the projects Terelis continued was the time travel Project which had been so helpful for him. Baltisar continued To help Terelis on all scientific projects. One additional note was the extra organ was used to receive and process Kurinite. The lack of this was responsible for their general lack of vitality. The natural foods help cure the sterilization of the Women as well.

The Council of the 12 which had new members, members Morlock, Lorbal Carmen Melonball were banned to the moon for life. The main goal would be survival, no time to poison the people. Terelis felt this was a suitable punishment. Paul and Bob were named honorary members of the council.

When Earth Matured a foundation of planets was formed.

This protected other worlds , but did not interfere with their

Development. This group Elected Fenusia to be the first

member Of this organization. Travelis Jr. and Angelica lived to

see this!

The End

Some dialog came from the movie "Daddy Long Legs" an MGM

Movie made in 1955

Star Trek Novel "Balance of Terror"

Star Trek Novel The Cloud Minders"

Star Trek Novel Plato's Step Children"